TREMBLE

A Child's First Book on Space

A Primer on the Fear of God

by R. A. Sheats

Second Edition

Formatter: Genevieve Sheats

Published by:

Psalm 78 Ministries
P. O. Box 950
Monticello, FL 32345
www.psalm78ministries.com

ISBN: 978-1-938822-48-3

Printed in the United States of America.

IN THE BEGINNING
GOD CREATED
THE HEAVEN AND THE EARTH. . . .

What does space look like?
What's out there?
And what does it teach us about God?

When you look up into the sky at night,
what do you see?

How did God make it all?
Who taught Him how to do it?

How wise must God be
to have created all these things?

GALAXY

A galaxy is a group of stars, dust, and
other matter held together by gravity.

There are over
170 galaxies
visible in this image alone.

"Who hath directed the Spirit of the LORD,
or being His counselor hath taught Him?
With whom took He counsel,
and who instructed Him?"

Isaiah 40:13-14

The two galaxies in this image appear to have been tangled together.

What caused this?
We don't know, but God does.

"The LORD is great, and greatly to be praised: He is to be feared above all gods. For all the gods of the nations are idols: but the LORD made the heavens. . . . O worship the LORD in the beauty of holiness: fear before Him, all the earth."

Psalm 96:4-9

Barred Sculptor Galaxy

This galaxy has an incredibly large star formation in its nucleus (its center).
You can see the star formation by its red hues.

"Among the gods there is none like unto Thee, O Lord; neither are there any works like unto Thy works."

Psalm 86:8

NEBULA

A nebula is a giant object in space believed to be formed of gas and dust.

This nebula is found within our own galaxy.

"By the word of the LORD
were the heavens made;
and all the host of them
by the breath of His mouth. . . .
Let all the earth fear the LORD:
let all the inhabitants of the world
stand in awe of Him."

Psalm 33:6-8

The Hourglass Nebula

Astronomers wonder about the unusual formation of this planetary nebula.

Pinwheel Galaxy

"Be wise now therefore, O ye kings: be instructed, ye judges of the earth. Serve the LORD with fear, and rejoice with trembling. Kiss the Son, lest He be angry, and ye perish from the way, when His wrath is kindled but a little. Blessed are all they that put their trust in Him."

Psalm 2:10-12

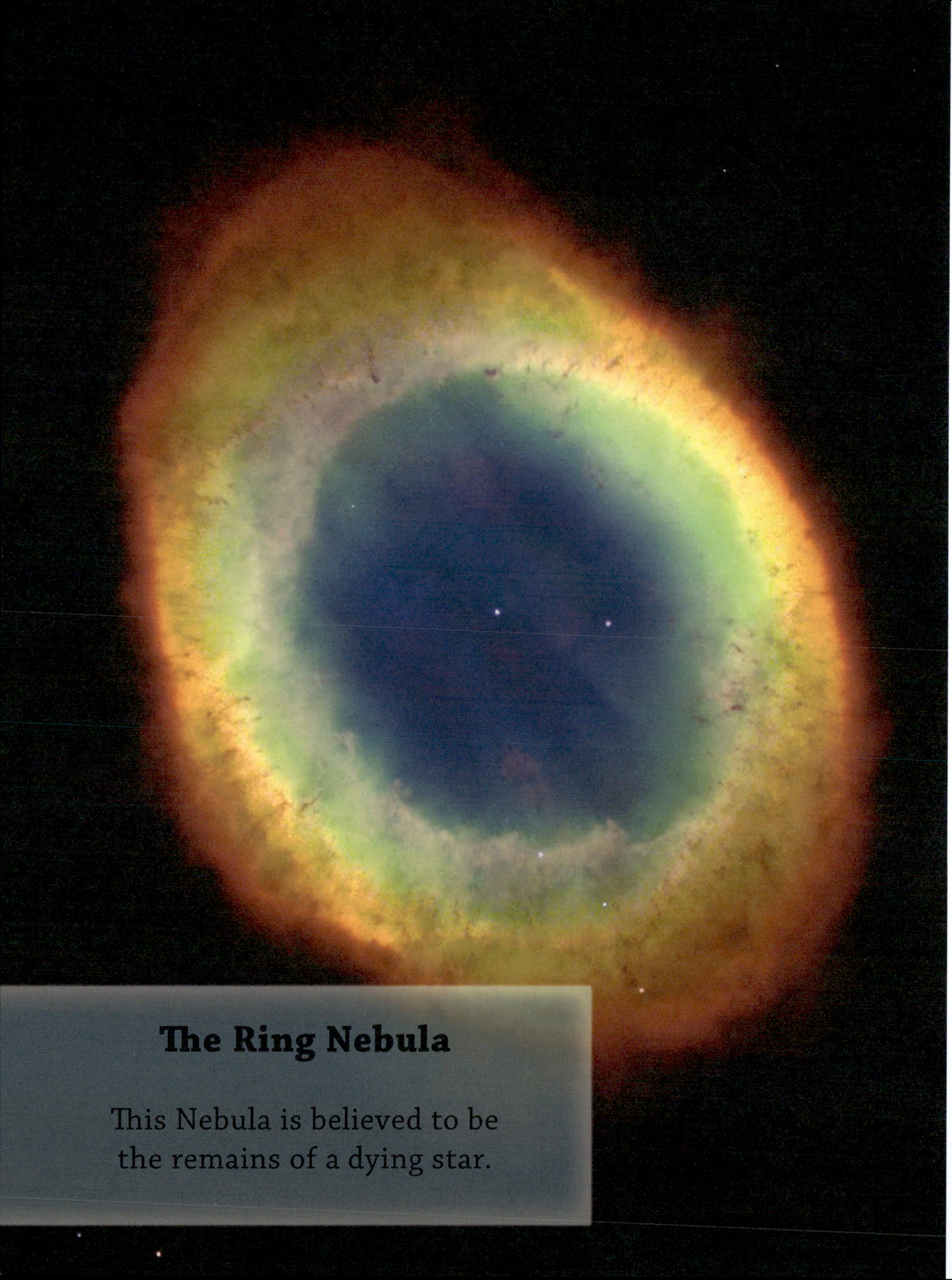

The Ring Nebula

This Nebula is believed to be the remains of a dying star.

"He stretcheth out the north over the empty place, and hangeth the earth upon nothing. . . . The pillars of heaven tremble and are astonished at His reproof."

Job 26:7, 11

Several hundred galaxies are visible
in this single image.
It is difficult for us to even count them all,
but God knows each one by name.

"Lift up your eyes on high,
and behold who hath created these things,
that bringeth out their host by number;
He calleth them all by names."

Isaiah 40:26

But the Lord is the true God,
He is the living God, and an everlasting King:
at His wrath the earth shall tremble, . . .
He hath made the earth by His power,
He hath established the world by His wisdom,
and hath stretched out the heavens
by His discretion."

Jeremiah 10:10-12

Can you make a
spiral galaxy like this?

Crab Nebula

This nebula is incredibly complex. It is known as one of the most intricately structured objects ever observed in space. How did God create this?

Cat's Eye Nebula

This nebula has been seen expanding
for 20 years.

"O the depth of the riches both of the wisdom and knowledge of God! How unsearchable are His judgments, and His ways past finding out! For who hath known the mind of the Lord? or who hath been His counsellor? . . . For of Him, and through Him, and to Him, are all things: to whom be glory for ever. Amen."

Romans 11:33-36

"The LORD reigneth; let the people tremble. . . . Let them praise Thy great and terrible name; for it is holy."

Psalm 99:1-3

This galaxy is believed to be 50,000 light-years across.

A light-year is the distance light travels in one year, nearly 6 trillion miles!

Three main galaxies are visible in this image.
The largest is thought to be 522,000 light-years across
from the tip of one arm to the tip of the other.

Can you count to 522,000? Can God count that high?

"Thou, even Thou, art LORD alone;
Thou hast made heaven,
the heaven of heavens,
with all their host, . . .
and Thou preservest them all;
and the host of heaven
worshippeth Thee."

Nehemiah 9:6

This interesting formation is a *dwarf irregular galaxy.*

This globular cluster of millions of stars is so bright that it can be seen with the naked eye.

"The heavens shall praise Thy wonders, O LORD: Thy faithfulness also in the congregation of the saints. For who in the heaven can be compared unto the LORD? . . . God is greatly to be feared in the assembly of the saints, and to be had in reverence of all them that are about Him."

Psalm 89:5-7

Dusty matter hides many brilliant stars from view in this nebula located in the constellation Orion.

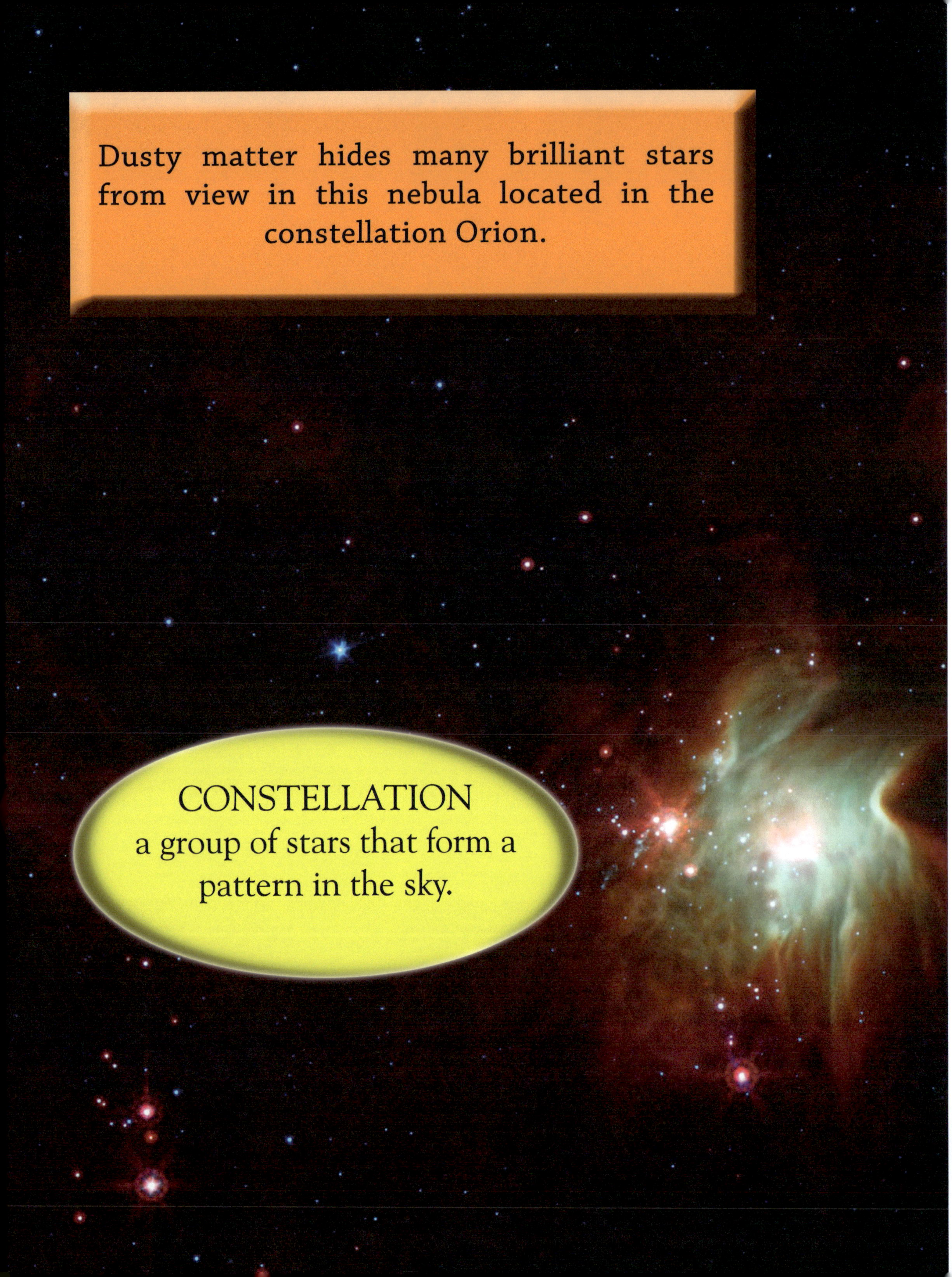

CONSTELLATION
a group of stars that form a pattern in the sky.

"Seek Him that maketh the seven stars and Orion,
and turneth the shadow of death into the morning,
and maketh the day dark with night. . . .
The Lord is His name."

Amos 5:8

This side view of a disk spiral galaxy reveals the beautiful dust and matter that makes up its small spirals.

The blue hue is made from stars of that color.

MONKEY HEAD NEBULA

"O LORD, how manifold are Thy works! in wisdom hast Thou made them all: the earth is full of Thy riches."

Psalm 104:24

Astronomers call this a “grand design” spiral galaxy because of its well-defined spiral motion.

This barred spiral galaxy is a beautiful display of the power and wisdom of its Grand Designer.

“Lo, these are parts of His ways;
but how little a portion is heard of Him?
but the thunder of His power
who can understand?”

Job 26:14

This nebula is located in Orion. It is a reflection nebula: it has no light of its own, but is seen by the bright star shining behind it.

The black area at center is believed to be a black cloud so dense that no light can penetrate it.

"Great and marvellous are Thy works, Lord God Almighty;
just and true are Thy ways, Thou King of saints.
Who shall not fear Thee, O Lord, and glorify Thy name?
for Thou only art holy."

Revelation 15:3-4

Within this image, every speck of light
(even the smallest) is a galaxy.
Because they are so far away,
even the brightest objects in this picture
cannot be seen with the naked eye.

How powerful must God be
to have made all of these?

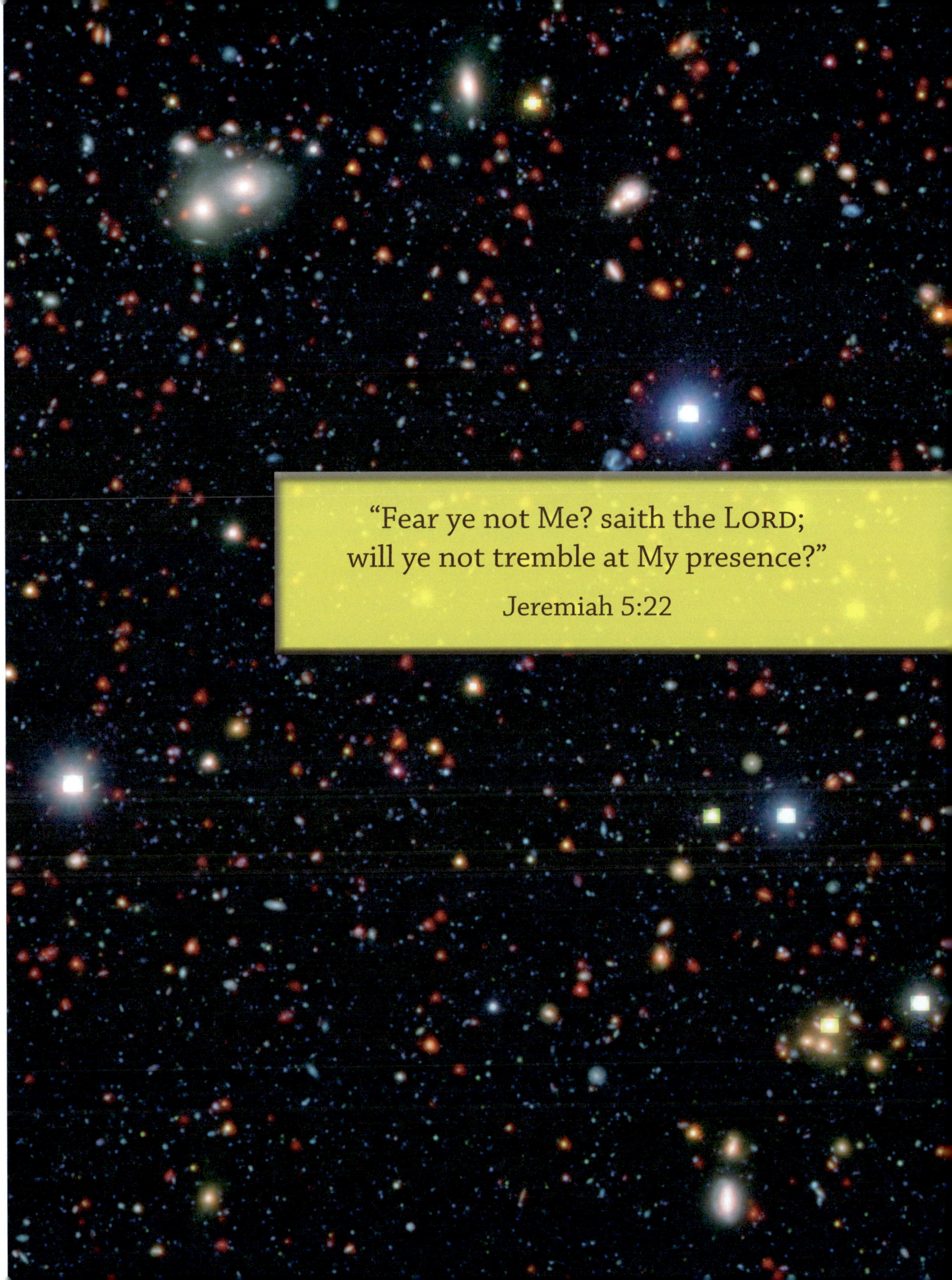

“Fear ye not Me? saith the LORD;
will ye not tremble at My presence?”

Jeremiah 5:22

"Thus saith the LORD, The heaven is My throne, and the earth is My footstool. . . . For all those things hath Mine hand made, and all those things have been, saith the LORD: but to this man will I look, even to him that is poor and of a contrite spirit, and trembleth at My Word."

Isaiah 66:1-2

These pillars of gas are found in the Eagle Nebula

All photos created by NASA, Hubble Heritage Team/JPL

Earth cover image: http://earthobservatory.nasa.gov/IOTD/view.php?id=885
NASA's Earth Observatory. Image created by Reto Stöckli, Nazmi El Saleous, and Marit Jentoft-Nilsen, NASA GSFC

Messier 104: NASA/STScl/AUPA

The Hourglass Nebula: NASA/JPL-Caltech/ESA, the Hubble Heritage Team (STScl/AURA)

PIA 14443 The Ring Nebula: NASA/JPL-Caltech, the Hubble Heritage Team (STScl/AURA)

PIA 13316 Tangled Galaxies: NASA/CXC/SAO/JPL-Caltech/STScl

NGC 1512: NASA/Space Telescope Science Institute

Globular Cluster: Credit:NASA/JPL-Caltech/NOAO/AURA/NSF

M16 Pillars of Creation: NASA/ESA/STScl/J. Hester and P. Scowen (Arizona state University)

PIA 16884 Small Mag: NASA/CXC/JPL-Caltech/STScl

PIA 17256 / NGC 253 Barred Sculptor Galaxy: NASA/JPL-Caltech

PIA 15630/M101 Pinwheel Galaxy Rainbow: NASA/JPL-Caltech/ESA/STScl/CXC

NGC 2818 Planetary Nebula:

PIA 16613: NASA's Goddard Space Flight Center/ESO/JPL-Caltech/DSS

Crab Nebula: NASA/ESA/JPL/Arizona State University

Cat's Eye Nebula: NASA,ESA, HEIC, and the Hubble Heritage Team (STScl/AURA)

I Zwicky 18: NASA, ESA, and A. Aloisi (Space Telescope Science Institute and European Space Agency, Baltimore, Md.)

NGC 5866 Disk Spiral Galaxy

NASA and the Hubble Heritage Team (STSci/Aura)

NGC 1999 Nebula in Orion: NASA and the Hubble Heritage Team (STScl)

Barred Spiral Galaxy NGC 1300

NGC 2174 Monkey Head Nebula: NASA, ESA, and:the Hubble Heritage Team (STScl/AURA)

PIA 03244 Christmas Tree Star Cluster: NASA/JPL-Caltech/Cfa

NASA/JPL/STScl Hubble Deep Field Team

NASA, The Hubble Heritage Team and A. Riess (STScl)

Made in the USA
Las Vegas, NV
10 November 2026

57845499R00026